The Ultimate Samsung Galaxy S9 & S9 Plus Manual:

Tips, and Tricks to Optimize your Samsung Galaxy S9 and S9 Plus

Olly J Hart

© Copyright 2018 by Olly J Hart - All rights reserved.

The following eBook is reproduced below with the goal of providing information that is as accurate and reliable as possible. Regardless, purchasing this eBook can be seen as consent to the fact that both the publisher and the author of this book are in no way experts on the topics discussed within and that any recommendations or suggestions that are made herein are for entertainment purposes only. Professionals should be consulted as needed prior to undertaking any of the action endorsed herein.

This declaration is deemed fair and valid by both the American Bar Association and the Committee of Publishers Association and is legally binding throughout the United States.

Furthermore, the transmission, duplication, or reproduction of any of the following work including specific information will be considered an illegal act irrespective of if it is done electronically or in print. This extends to creating a secondary or tertiary copy of the work or a recorded copy and is only allowed with the express written consent from the Publisher. All additional rights reserved.

The information in the following pages is broadly considered a truthful and accurate account of facts and as such, any inattention, use, or misuse of the information in question by the reader will render any resulting actions solely under their purview. There are no scenarios in which the publisher or the original author of this work can be in any fashion deemed liable for any hardship or damages that may befall them after undertaking information described herein.

Additionally, the information in the following pages is intended only for informational purposes and should thus be thought of as universal. As befitting its nature, it is presented without assurance regarding its prolonged validity or interim quality. Trademarks that are mentioned are done without written consent and can in no way be considered an endorsement from the trademark holder.

Table of Contents

Introduction   7

Chapter 1: Setting Up the Phone   9

    Activate the Phone   9
    Restore From Backup   12
    Transferring Data from another Phone   12
    Security Settings   15

Chapter 2: Learning Tips and Tricks   18

    Activate Fingerprint Sensor Gestures   18
    AR Emoji   18
    Bixby   19
    Bixby Vision   20
    Mastering the Camera   21
    Picture Size   21
    Rear Camera   22
    Conserve Battery Life   23
    Change Installed Keyboards   24
    Customize the Navigation Bar   25
    Display Settings   25
    Dual Audio/Bluetooth   26
    Find My Mobile   27
    Google Play Store   27
    Schedule the Blue Light Filter   28
    Split-Screen Mode   28
    Touch Sensitivity   29

Chapter 3: Best Essential Apps — 30

Adobe Acrobat Reader — 30
Android Auto — 30
CamScanner — 31
Clean Master - Space Cleaner and Antivirus — 31
Facebook Messenger — 33
Feedly — 33
Gboard — 33
Gloud Games — 35
Google Duo — 35
Parallel Space - Multiple Accounts and Two-Face — 35
PPSPP - PSP Emulator — 36
Power Director Video Editor App — 36
Samsung Max — 37
Samsung Smart Switch — 38
Spotify — 38
Trello — 39
Unified Remote — 39
VLC for Android — 40
Xender — 40

Chapter 4: Common Problems & Solutions — 41

Apps Not in Full Screen — 41
Connectivity — 41
Dark Slow Motion Videos — 43
Dead Zone on Screen — 44
Edge Lighting Not Working — 44
Keyboard not Appearing — 46
Messaging App Sending Photos Automatically — 46
Missing App Drawer — 47

Out of Storage	47
Samsung Pay	49
Takes Too Long to Charge	49
Video Stutters When Recording in 4k	50
Wet Speaker Audio Problems	51

Chapter 5: S9/S9 Plus Maintenance

53

Antivirus	53
Battery	53
Backup	54
Device Maintenance App	56
Reset	57
Memory	58
Performance	58

Conclusion   60

Introduction

Congratulations on purchasing ***Ultimate Samsung Galaxy S9 & S9 Plus Manual: Tips, Tricks, and Great Optimization for Your Samsung Galaxy S9***, and thank you for doing so. This book will help you become familiar with your Samsung Galaxy S9 or S9 Plus. After reading this book, you will know the ins and outs of your phone, be knowledgeable about your phone, and be able to solve problems that confuse many users.

The following chapters will discuss a variety of topics important to owning a Samsung Galaxy S9 or S9 Plus. The S9/S9 Plus is one of the best phones currently available, and it is Samsung's flagship product on the smartphone market. Inside this book, you will find information on how to set up your S9/S9 Plus, tips and tricks for using and customizing your phone, the best apps to use with your phone, solutions to common problems with the S9/S9 Plus, and how to perform routine maintenance on your phone. Armed with the knowledge, in this book, you will be a whiz at configuring and navigating your phone as well as solving common problems that you might face. This book is good for you no matter if you are just starting out or if you have had your phone for a while.

This book assumes that you have a basic understanding

of the Galaxy S9 or S9 Plus. Before reading this book, you should be familiar with how to tap, click, press, and use your finger as well as general navigation of your phone. Once you have these basic skills, it is time to learn more about your phone and to take your S9 skills to the next level.

There are plenty of books on this subject on the market, so thanks again for choosing this one! Every effort was made to ensure it is full of as much useful information as possible. Please enjoy!

Chapter 1: Setting Up the Phone

Activate the Phone

The first thing to do is take the Samsung S9/S9 Plus out of the box and insert the SIM card into the phone. Hold the Power button down to power the phone on. The phone is on when the Samsung Galaxy S9 screen appears. Once it is on the screen, release the Power button.

Next, select the language of your choice and press the **Arrow** on the right. It will take a few seconds to complete the activation.

Once the activation is complete, the phone will allow you to connect to Wi-Fi. Choose the Wi-Fi network you want to use and enter the Wi-Fi password. Once you are connected, the phone will prompt you to download any updates that are available. It is important to install any updates for the phone, as they may be very important and might contain security features needed to keep the phone secure.

Next, the Terms and Conditions screen will appear. Select each of the documents and then tap **Agree** to continue. There is an option for **Agree to all**. The documents are the Terms and

Conditions, the Privacy Policy, and the Diagnostic data agreement. You should review each document before agreeing.

With some carriers, there may be a screen with data regarding the phone plan. Not all carriers have this screen appear.

On the next screen is a Google account login page. Put in your Google email or phone number. There is an option to skip this, however, not using your Google account will not let you access all of the Google features on the phone, so you should enter your Google information. If you do not have a Google account you can click on **Create account** to make a new Google account.

Once you have added your Google account, it will show you any backups you have to choose from. If you have no backups or if you want to start the phone with factory settings, press **Don't Restore**. It will take a few seconds for the phone to activate.

Next, you have the option to set up the security settings including Intelligent Scan, Face Recognition, Iris Scanner, Fingerprint Scanner, or PIN, pattern, or password. There is a **No, thank you** option. We will discuss the security settings later in this chapter.

The following screen will be a Google services page. You can select or adjust any of the Google settings then press **Agree**. Review any other apps to be downloaded and press **Ok**.

It will then display the Google Assistant page. Follow the commands to set up Google Assistant.

Next, either sign in or create a Samsung account. You can bypass this by pressing **Skip** and **Skip** again. The screen will show the Samsung Terms and Conditions. Review that document and then tap **Agree**.

The Restore Data screen will appear. You can select to restore data from the Samsung cloud, from your old device, or not restore at all, and set up the phone as a new device. Then tap **Next**.

The phone will finish activating and the phone carrier may have additional Terms and Conditions to agree to. This process can vary depending on carrier.

Finally, there is an **All Done** screen and the phone will show the Home screen.

Restore From Backup

On the Home Screen, place your finger in the center of the display and swipe either up or down. This will open the apps screens. Navigate to **Settings**, then click **Accounts,** and tap **Backup and restore**. Select Google accounts, swipe the **Back up my data** switch into either on or off. Select the backup from a list. If you need to change the Backup account field, tap Backup account, select the account, and tap Automatic Restore.

Transferring Data from another Phone

There are several methods of transferring data from your old Android or iPhone that require the use of the Samsung Smart Switch app. You can download the Samsung Smart Switch app from the Google Play store. While some of these techniques are unique to iPhones, the majority of them will work for transferring from an Android phone as well.

To transfer the data from an iPhone wirelessly to your S9, you need the Samsung Smart Switch app, which can be downloaded from the Google Play store. Using the app, you can access backups from your iCloud account. This method of transferring data will only transfer your Calendar, Contacts, Videos, and Photos, and this method is only partially supported.

The first step is to back up your iPhone to your iCloud account. The next step is to open the Smart Switch app on your

Samsung S9/S9 Plus. First, touch **Wireless** and touch **Receive** and choose **iOS**. You will be prompted to enter your Apple ID and password. Next, touch the **Sign in** button. Select the basic content you want to import from iCloud then press **Import**. There are two pages where you should touch **Import**. You can also do this step by choosing Android instead of iOS.

The next method involves the use of a USB or Lightning cable using the Samsung Smart Switch app. First, launch the Samsung Smart Switch app on the Samsung phone. Next, you need to attach the USB or Lightning cable to the Android or iPhone and connect the mini USB to the S9/S9 Plus. A message will appear on the iPhone asking if you want to trust the device. When it appears tap **Trust** and then touch **Next**. Select the data that you want to transfer and then tap **Transfer**.

Another method involves the use of the **Syncios Data Transfer application** for your computer. There is both a Windows and a Mac version of data transfer software. Download the application to your computer and launch it. Three options will appear. They are Transfer, Restore, and Backup. Select **Transfer**.

Connect both phones to your computer with a USB or Lightning cable. In order for the Samsung phone to appear correctly in the Syncios Data Transfer application, you may need to enable USB debugging mode on the Samsung phone. To do this, you first need to enable Developer Options on your phone. Instructions on how to access the Developer Options is in **Chapter 4.**

If transferring from an iPhone, a pop-up message will appear. Click **Trust** or **Allow** on the iPhone. It is important that the Samsung phone should be on the right and the iPhone is on the left of the screen. In order to change the position of the phones, click the Flip option located in-between the icons of the two phones. Click **Next** to select the content you want to transfer and then click **Next** to begin the transfer. Wait for the transfer to complete.

Another method also uses the Syncios Data Transfer app. This method uses an **Android** or **iTunes** backup made on a computer instead of backups synchronized with iCloud or another cloud storage service. Launch Syncios Data Transfer and navigate to the Restore mode and click on the Android or iTunes backup. Connect your S9/S9 Plus to the computer. Syncios will detect any Android or iTunes backups and will display them on the left panel. Choose a backup. Finally, select the contents from the middle box of the iTunes backup. Tap on **Start Copy**. Just wait while Syncios transfers the data.

The last way to load your iPhone backup onto your S9/S9 Plus is to transfer the data directly without having the iPhone present. You should back up the iPhone before starting. Open the Syncios Data Transfer application and choose **Restore** and select **iCloud Backup**. Connect the Galaxy S9 to the computer and log in with your iCloud Apple ID. Next, you can select the backup file you want to load and choose the content you want to load. Click **Next** to being the transfer.

Security Settings

The Samsung Galaxy S9/S9 Plus has several security features that allow you to access the phone. These are Intelligent Scan, Face Recognition, Iris Scanner, Fingerprint Scanner, and PIN, pattern, or password. Below are instructions on how to set up each security feature:

- Intelligent Scan

To activate Intelligent Scan, open the Settings app. Next, click on **Lock Screen and Security**. Navigate to Biometrics and select Intelligent Scan. If you have not done so, the phone will prompt you to set a pin number. The verification information screen will appear. Tap **Continue** to go onto the next screen, which lets you register your face. The phone needs to be 8 to 20 inches away from your face. Next tap **Continue**. Adjust the phone so your face is centered in the circle on-screen.

It is now time to register your irises. Remove your contacts or glasses and press **Continue**. Make sure that your eyes match the circles on the screen.

Once you have registered your irises and your face, turn on **Intelligent Scan**. Be sure to have the Intelligent Scan unlock and Screen-on Intelligent Scan toggled on.

- Fingerprint

To add your fingerprint, navigate to the **Settings App**. Click on **Lock screen and security**. Tap on **Fingerprint Scanner** and follow the instructions that appear on-screen. Place your finger over the fingerprint sensor. You can add additional fingerprints by clicking **Add**. It is recommended that you test the fingerprint sensor to make sure it is working. Next, make sure that fingerprint unlock is turned on. If not already, then tap to the **Turn on** position.

- Pin, Pattern, or Password

To add a Pin, Pattern, or Password to the phone, navigate from the Home screen to the Settings app. Locate **Lock screen and security**. In the Phone Security, choose Screen lock type. Choose one of the following: Swipe, Pattern, PIN, Password, or None. Choose the one you want to set-up.

To set up a Pattern, draw an unlock pattern that connects at least 4 dots. Once done, tap **Continue**. Draw the same pattern again and click on **Confirm**. You may be presented with a Notifications switch on the upper right. Make sure this is turned on and click on **Done**. There are some customization options that you can turn on including Hide content, Notifications icons only, Transparency, and Auto-Reverse text color.

To set up a PIN number for the device, enter a numeric PIN that is 4 to 16 digits long and tap Continue. Re-enter the PIN and then tap **Ok**. The rest of the steps are the same as the Pattern set-up.

To set up a password, enter the password. The password must be 4 to 16 characters long and must include one letter. Once entered, tap **Done**. The remaining steps are the same as the above set-ups.

Chapter 2: Learning Tips and Tricks

There are many tips and tricks for even the most advanced user of the Samsung Galaxy S9/S9 Plus. Below are several tricks for the phone.

Activate Fingerprint Sensor Gestures

You can activate the Fingerprint sensor gesture to swipe down the notification bar. This allows easy access to the notifications bar. It is especially useful on the S9 Plus. With the larger screen on the S9 Plus, it is sometimes difficult to reach the top of the screen. Meanwhile, the fingerprint sensor is easy to operate. To activate the fingerprint sensor for gestures, first, navigate to **Settings**, then **Advanced features**. Select **Fingerprint sensor gestures** and toggle the switch to on.

AR Emoji

The AR Emoji is an emoji that you can set up that will mimic the movements of your face. This feature allows the user to create a cartoon of themselves using the camera. To create an AR Emoji, navigate to the **Camera** app and make sure the front-facing camera is on. There are options on the top of the screen. You can either swipe or tap on **AR Emoji**. At the bottom of the screen is a face symbol. Tap it and then tap **Create**

Emoji. Remove any glasses, make sure that hair does not block your face, and smile with your lips closed. The app will prompt you for your gender. Afterwards, it will create the AR avatar and you can customize the emoji by changing skin tones, adding glasses, hair color, and select an outfit. When it is completed, the emoji will appear in the camera app below the viewfinder.

To use the AR Emoji, navigate to the Messenger app and tap the Sticker button. You can also access the AR Emoji from the Camera app. When in the Camera app, tap the **AR Emoji,** and you can create videos and pictures that will mimic the expressions on your face.

Bixby

Bixby is the Samsung AI assistant built into the S9/S9 Plus phone. Like Google Assistant, Bixby has a large database of information like restaurants and maps. The advantage Bixby has is that it was built specifically for Samsung phones by the manufacturer, so it may be able to perform tasks that Google Assistant can't handle.

You can access Bixby in two ways, either by giving it a voice command or by pressing the dedicated Bixby button on the side of the phone. Before using Bixby, however, it is important to register your phone with the AI platform so that Bixby learns and improves itself while you use it.

The first thing to do to set up Bixby is to tell it what language to use. Bixby is available in English, Korean, and Mandarin. After selecting the language, test Bixby to make sure that it understands your voice.

Bixby is useful as a search tool and can retrieve information quickly. Because it is built specifically for Samsung phones, Bixby can control a lot of actions without you having to press a lot of buttons, which makes Bixby good for hands-free.

Bixby Vision

Bixby has visual applications as well. To access Bixby Vision, navigate to the Camera app and launch it. Press the **Bixby symbol** to enable Bixby Vision. The symbol looks like a stylized eye.

Bixby Vision includes several new functions such as translating languages, the ability to scan QR codes, information about wines, transforming images into text, make-up recommendations, shopping mode, identifying objects, and calorie counting.

A few options make Bixby even more useful. You can enable Voice Wake-Up so that you can call to your phone without holding it. To activate this feature, navigate from **Bixby Home** and click the **menu overflow icon**. Next, tap **Settings** and then tap **Voice wake-up**. Enable voice wake-up by swiping from off to on.

Next, navigate to the **Bixby Voice prompt** and tap the **Bixby logo**. Follow the voice prompts and after reading each voice prompt, press **Done**.

You can also set Bixby for push-to-talk, which makes using the dedicated Bixby button more effective because it is waiting for the voice command instead of going immediately to Bixby Home. This can be enabled by pressing the **menu overflow icon**. Navigate to the General tab and click **Bixby key.** Next tap **Don't open anything**.

Mastering the Camera

There are many settings for the phone's cameras, including both the back and front cameras.

Picture Size

The first setting you should know about your camera is how to control the picture size. Note that picture size refers to the resolution of the picture and not the size of the picture file. The picture size is broken down into the number of pixels and the aspect ratio. The camera can shoot pictures from 3.7 MP to 12 MP. The aspect ratio's that are available are: 4:3, 16:9, 18.5:9

and 1:1.

Rear Camera

The S9 sports a rear camera that is a super speed dual pixel camera with 12MP AF sensor and pixel size of 1.4μm. The sensor ratio is 4:3. The S9 Plus has two rear cameras, a wide-angle camera, and a telephoto camera. The wide-angle camera has the same specifications as the camera on the S9. The Telephoto camera also has a 12MP AF sensor, the pixel size is 1.0μm, and it also has a sensor ratio of 4:3.

The S9's default picture size is 12 MP in 4:3 aspect ratio. The other picture sizes are 1:1 for web graphics; 16:9, which is good for TV and smartphone displays; finally there is 18.5:9, which is good for phones with 18:9 or 18.5:9 aspect ratio. Please note that the default picture size of 4:3 is the best for print photos.

Typically, it is best to use the larger pixel size. Pictures taken with higher numbers of pixels can be scaled down to lower resolution without degrading the image, but a picture taken with a lower number of pixels does not look good when it is scaled up.

There are three things to consider when choosing which video size (size of the file) to use. The first is the frame rate. Normal speed videos, typically shot at 30 fps (frames per second), can be shot at up to 60 fps. You can record FDH 1080p for slow motion (240 fps) or HD 720p (960 fps) for super slow-motion. Next is the aspect ratio. There are 3 aspect ratios available for video: 16:9, 18.5:9 and 1.1. Finally, there is video size (resolution). The S9/S9 Plus can record videos with up to 4k UHD resolution.

Conserve Battery Life

Every phone has battery life issues at some point. Here are some tricks to conserve your battery for longer and healthier battery life.

First, do not increase the screen resolution on your phone. There are three settings for screen resolution. They are HD+ (1480x720), FHD+ (2220x1080), and WQGD+ (2960x1440). The larger the resolution, the more battery life that is used. You can locate the resolution settings by going to **Display**, navigating to **Screen resolution**, and using the slider on the right to choose your resolution.

The next method for conserving battery life is to put apps to sleep permanently. You can navigate to **Device maintenance** and press **Optimize now**. The phone will display a list of apps that are using storage. This will free up storage space.

Another method of conserving power that involves putting apps to sleep is to navigate to **Device maintenance** and select **Battery**. Under battery, there are options for power saving modes. There is also an App power monitor. The App power monitor shows a list of apps that are using power. You can select which apps you want to be turned on and then press **Save power**. Be aware that an app that is put to sleep does not send notifications to the phone, so apps you want notifications from should be unchecked.

The next method of conserving power is to disable the Always On Display. The Always On Display updates with notifications even if the screen is off, however, it uses a lot of power and can quickly drain your battery. To turn the Always On Display off, navigate to **Settings** and select **Lock screen and security** then slide the **Always On Display** slider into the off position.

Change Installed Keyboards

The S9/S9 Plus allows the user to choose which keyboard layout they would like to use. You can install new keyboard layout apps from the Google Play store. One popular keyboard is Gboard, which we will discuss in Chapter 3.

In order to change the keyboard, navigate to **Settings**, click on **General management**, select **Languages and**

input. From there, you can choose which keyboard that you want to use.

Customize the Navigation Bar

The Navigation bar on the bottom of the screen is in a different order from other Android phones. The S9/S9 Plus has Recent — Home — Back, while most Android phones have Back — Home —Recent. If you are a long time user of Android phones, it can be challenging to have something so basic altered. There is, however, a way to customize the Navigation bar.

To change the Navigation bar, navigate to **Settings** and tap on **Display**. Select **Navigation bar** and click on the order, and it will bring up a pop-up where you can choose the order you want the Navigation bar to appear in.

Display Settings

To adjust the display Settings for the phone, navigate to **Settings** then select **Display**. There are four options under Display. They are Blue light filter, LED indicator, Block accidental touches, and Screensaver. You can turn these settings on or off by tapping the slider into the on or off position.

As discussed above, you can alter the screen resolution and the Always On Display from **Display** and selecting the option you want to change.

You can also change the screen mode. The screen mode optimizes the screen for certain types of content. Each mode is designed for a specific reason. There are four options: Adaptive display, AMOLED cinema, AMOLED photo, and basic. The Adaptive display optimizes sharpness, color range, and saturation. The AMOLED cinema mode is made for viewing videos. The AMOLED photo is great for viewing pictures, and the basic mode is excellent for daily use and does not include changes to the picture quality, size, or other adjustments. Third-party apps may not work with Adaptive Display.

Dual Audio/Bluetooth

One fascinating feature of the S9/S9 Plus is the phone's ability to handle two Bluetooth audio signals, each from a different device. For example, you can connect to two headphones at the same time, allowing two people to listen to the same music or audio. This feature especially comes in handy if you are traveling with children. While traveling, you can have one screen with two headsets, allowing the children to watch the same show on the phone without causing a distraction.

The first step in using Dual Bluetooth is to pair the headsets to the phone. First, you need to make sure that Bluetooth is enabled on the S9/S9 Plus. From the top of the Home screen, swipe down to reveal the **Notification Shade**.

Tap the **Bluetooth symbol** to activate Bluetooth.

To pair the devices, swipe from the top of the screen again and reveal the **Notification Shade**. Tap on **Bluetooth** and go to **More Settings**. Select the name of the Bluetooth headset to pair it.

Next, you need to enable Dual audio. Navigate to **Settings**, click on **Connections**, and tap **Bluetooth**. On Bluetooth, tap the **three-dot icon** and click on **Dual audio**. Slide the toggle to activate the dual audio feature.

Find My Mobile

You can improve the performance of your Find my mobile feature. Changing some of the settings makes it easier for the phone to keep track of your current location. Accurate location information makes it easier to find a lost phone. In order to improve Find my mobile, navigate to **Settings**, next select **Lock screen and security**, then click on **Find my mobile**, and finally swipe the toggle for **Send last location**.

Google Play Store

The Google Play Store is an app marketplace where you can download and purchase new apps for your phone. The Google Play store is accessible either from your phone or on the web. To access the marketplace on your phone, look in the app drawer on the Home screen or in apps. Once you navigate to the store, load the app. Next, locate an app that you want to download. You can either search or browse through the apps in order to find the app you are looking for. Apps are either purchased or downloaded for free. Once you have located the app, you need to install it. Navigate to the description screen and press **Install**.

Schedule the Blue Light Filter

The Blue light filter makes it easier to see the screen in the dark, and it causes less strain on the eyes while doing so. The S9/S9 Plus comes with a feature that automatically turns on the blue light filter at a specific time so that as it gets darker, you can have an easier to read phone without having to always change display settings.

To schedule the blue light filter, navigate to **Settings**, go to **Display**, select the **Blue light filter** and swipe **Turn on as scheduled**. You can then choose either the default of **Sunset to sunrise** or you can make a custom schedule.

Split-Screen Mode

Another useful feature of the S9/S9 Plus is the ability to turn on Split-screen mode, which allows you to be in two apps at once. It is very useful to have split-screen set up with your Recents button, which is the default for Android phones, but the feature is disabled in the factory settings. To turn on this useful feature, navigate to **Settings**, click on **Advanced features**, then **Multi-window**, and toggle the switch for **Use recent button**.

Touch Sensitivity

Another good tip is to change the touch sensitivity on the S9/S9 Plus. Changing the touch sensitivity comes in handy when you are using a protective case or screen protector. You can enable touch sensitivity by navigating to **Settings** and select **Advanced features**. From there select **Touch Sensitivity** and slide it into the on position.

Chapter 3: Best Essential Apps

There are many apps for Android and the S9/S9 Plus in particular. Below are some of the best compatible and essential apps:

Adobe Acrobat Reader

Adobe Acrobat Reader is a powerful tool for viewing, reading, and editing PDF files. This app connects to your cloud storage so that you do not need to store the PDF files on your phone, which conserves storage. Adobe Acrobat Reader not only allows you to read PDF files, but it also allows you to edit them. This is useful for adding a digital signature or a filling out a form-fillable PDF.

Android Auto

Android Auto is a Google-made app designed for when you are driving. Android Auto features the Google Assistant for voice recognition to minimize distracted driving. It also allows easy access to music, GPS maps, and messaging apps. It also features larger buttons than usual, which makes it easier to manipulate the phone.

CamScanner

CamScanner is an app that lets you make digital copies of your documents. This app makes it easy to carry all of your important documents without having to carry those papers with you. The app uses the phone's camera to scan the documents and capture them as images. CamScanner can save the documents in a variety of file formats like Jpeg or PDF. The app makes accessing your documents very easy.

Clean Master - Space Cleaner and Antivirus

Clean Master is an excellent maintenance app. It has many features that help keep the phone working smoothly. Clean Master not only provides an antivirus, but it also features a junk cleaner, a battery saver, security features, a game enhancement feature, and other options.

Clean Master has the following features:

- **Junk Cleaner:** The junk cleaner helps the phone by removing residual junk, and cleaning cache files for social media apps without deleting the wrong files.

- **Antivirus:** An antivirus is an app that locates, blocks, and removes viruses and malware from the phone. This helps keep the phone in good working order.

- **Family Locator:** Use the family locator to track the GPS locations of your family members so that everyone knows where each other is. This is particularly useful for parents whose children have Samsung S9/S9 Plus phones.

- **Private Photos:** This feature of Clean Master allows you to encrypt your photos so that hackers cannot easily access them.

- **Wi-Fi Security:** The Wi-Fi security feature adds the ability to detect unauthorized connections and fake Wi-Fi signals.

- **Boost Mobile:** The Boost Mobile feature speeds up your phone by freeing up RAM.

- **Battery Saver:** The Battery Saver mode allows you to extend your battery life by hibernating apps that run in the background.

- **Game Master:** In Game Master mode, you can find and manage your games including controlling the load time for various games.

Facebook Messenger

Facebook Messenger is the messenger app associated with Facebook. It allows you to contact anyone on your Facebook Messenger contact list either by phone, video call, or instant message. Many users have Facebook Messenger even if they do not use Facebook.

Feedly

Feedly is an RSS reader that allows you to read blogs, view YouTube videos, read articles, learn new topics, track trends, and read magazines. Feedly is a well-designed app that gives you the content you need quickly so that you can stay abreast of the news, articles, and trends that you follow. With Feedly, you can add any RSS feed manually or you can browse articles and magazines you want to follow.

Gboard

Gboard is Google's alternate keyboard for Android devices. Gboard works well with the Samsung S9/S9 Plus, and it features several types of input. The keyboard includes the "G" button, which makes it easy to enter a Google search.

The features include:

- **Glide typing:** Glide typing is where you slide your finger from letter to letter until the word is spelled.

- **Voice typing:** Voice typing is an input where you dictate to the keyboard.

- **Handwriting:** Handwriting allows you to write in cursive in over 100 languages

- **Search and share:** Press the G button to activate the search and share option on the phone. This feature allows you to use Google search features.

- **Emoji search:** The emoji search allows you to locate emojis quickly and easily.

- **GIFs:** You are able to locate and share GIFs.

- **Multilingual typing:** Multilingual typing automatically locates which language you are writing in from the languages that are enabled.

- **Google Translate:** Google Translate translates the words that you are typing as you are entering it into Gboard.

- **Hundreds of language varieties:** There are literally hundreds of languages that you can use with Gboard. Go to https://goo.gl/fMQ85U for a full list of languages.

Gloud Games

Gloud Games is a gaming app that lets you play PC games on your phone. GloudGames is the perfect app for the PC gamer in you. The app has a large library of games that you can download and play.

Google Duo

Google Duo is among the best video calling apps available on Android. It has a simple user interface and is useful to keep in touch with distant people. Google Duo has the ability to make voice calls as well.

Parallel Space - Multiple Accounts and Two-Face

Parallel Space is an app that allows you to have multiple accounts for various social media apps on your phone. For instance, you can log into two different Facebook accounts at the same time. This helps when you have a business account and a personal account. You can switch between accounts with a tap. Parallel Space also allows you to create a unique space with themes. Themes can be switched between with a tap. Parallel Space can help with your phone's security by hiding apps that have sensitive personal information.

PPSPP - PSP Emulator

PPSPP is an emulator program that lets you play PSP games on your Android phone. Not only does PPSPP run older PSP games, but it also hosts many homebrew games. Homebrew games are games designed by fans as opposed to a game studio or a professional game designer. Games do not come with the app but you can create your own by making .CSO or .ISO files from the games you own. You can then load them onto the phone and play them in HD. You can also download games off the internet but those may be subject to copyright laws and downloading them may be illegal. There is both a free and a pay version of PPSPP.

Power Director Video Editor App

PowerDirector Video Editor is a great video editing app, which is a must-have when your phone's camera is as good as the S9's. PowerDirector allows even the newest user to edit and create videos easily. The PowerDirector app is based on a Windows application of the same name. PowerDirector has built a name for itself as easy to use software that makes excellent videos. Now, you can use the mobile version of the software to edit the impressive 4k resolution videos that are at your fingertips.

Samsung Max

Samsung Max is a powerful privacy protection and data management app. Samsung Max not only monitors and creates reports on your data usage, but it can also increase your security when connecting to Wi-Fi networks. Below are the features of Samsung Max:

- **Savings reports:** Savings reports provide you with information about the data your apps use and how to save data. Saving data helps keep you saving money on your data plan.

- **Manage apps:** You can control the data consumption of various apps and control how much data the apps are able to use.

- **Wi-Fi security:** Samsung Max encrypts all connections to Wi-Fi to help keep your data safe.

- **Incognito:** Incognito mode helps fight data snoopers and targeted ads by making it more difficult for the companies to track your usage.

- **Privacy reports:** Privacy reports let you know how to manage your privacy with regard to network connections and apps.

- **Boost Wi-Fi:** Boost Wi-Fi helps improve poor or weak signal areas.

- **Ultra apps:** Ultra apps are apps that are designed for Samsung's Max cloud technology, which optimizes the web apps, social media accounts, and other destinations. Ultra Apps helps protect your privacy and save your data.

Samsung Smart Switch

Samsung Smart Switch is a powerful app that helps you move your data from one phone to another. Not only does Samsung Smart Switch move files from various devices, but it can also create backups of essential data. More information can be found on Samsung Smart Switch in Chapter 1.

Spotify

Spotify is a popular music streaming app. It puts a large catalog of music of all genres into your hands. With Spotify, you can make playlists of your favorite music. Spotify free lets you play any playlist, artist, or album on shuffle mode. Spotify Premium also lets you play any song any time on any device, and the music is not shuffled. Premium members can also download music for the offline play and do not see or hear any ads.

Trello

Trello is a flexible to-do and organizational app that lets you organize whatever project you are working on. It allows you to use the app by yourself or with coworkers, family, and friends. With Trello you can customize your different workflows, add to-do checklists, assign tasks, attach files from Dropbox or Google Drive, reply to comments using Android Wear, upload videos and photos, and work offline. After working offline, the app will automatically sync when a connection is available.

Unified Remote

Unified Remote is an app that allows you to control your computer remotely. The app works with all types of computers from Windows machines to Mac and even Linux. Unified Remote is designed to work with over 90 compatible programs including mouse and keyboard control, media players, screen mirroring, file managers, and terminal access. There is both a paid and a free app. The paid app has additional features including 90 remotes, voice commands, and Android Wear (quick actions for both voice and mouse).

VLC for Android

No list of Android apps would be complete without VLC for Android. VLC is a media player that plays videos and music. Unlike other video players, VLC is known for being able to play even the strangest and rarest of video and music file formats. With VLC, you know you can view or listen to any such files. VLC is not just for Android but is available for a whole host of devices from Windows, Macs, iOS, and, of course, Android devices.

Xender

Xender is a music and file sharing app. It allows you to quickly transfer files to or from your phone. Xender supports Android, Tizen, PC to Mac cross-platform transferring, and Windows. Xender allows you to send any kind of file quickly between devices. This includes apps, photos, videos, and other files. Xender uses Bluetooth to do the transfers so there is no need for a cable or data usage. It also allows you to send large files, and you can view applications from friend's mobile apps. Xender also includes a powerful file manager to help you stay organized.

Chapter 4: Common Problems and Solutions

There are many common problems that you may experience with your S9/S9 Plus. Below are some details about common problems and their solutions.

Apps Not in Full Screen

Some older apps are not built for the current 18:9 aspect ratio. Instead, they are designed for a 16:9 screen. The S9/S9 Plus, however, has a solution. Navigate to **Settings** and select **Display**. Navigate to **Full screen option**. Make sure that all of the apps are using 18:9 ratio. Once they are all using the 18:9 aspect ratio, all of the apps will appear in full screen.

Connectivity

Cell phones constantly face the battle of good connectivity and being connected to a Wi-Fi or Bluetooth connection. We will cover both types of connectivity issue.

- **Bluetooth**

The first type of connectivity issue is with Bluetooth. If you are having issues with Bluetooth devices not connecting properly, the first troubleshooting step is to turn both devices off and then turn them back on. This is referred to as power cycling. Delete the Bluetooth profile for the Bluetooth device. Power the devices back on and pair back the two devices. The issue should now be resolved.

Sometimes, the phone can get clogged with old Bluetooth cache data. In order to clear the cache, navigate to **Settings** then click **Apps**, press the **three-dot button**, and scroll down to **Bluetooth** and push **Clear cache**. This will erase the previous history of Bluetooth devices.

- **Wi-Fi**

The second type of connectivity issue is with Wi-Fi. If you are having issues with your Wi-Fi, the first step is to forget your Wi-Fi network. To forget the network, navigate to **Settings** then **Connections** and then **Wi-Fi**. Once there, select the Wi-Fi connection and press **Forget**. This will make the phone delete the network connection. Once the connection is deleted, select the Wi-Fi signal again and re-enter the password to reconnect. If the phone continues to have problems, then the router may need to be reset. Every router, the hardware that gives you internet, has different settings, but typically, you just unplug the router for up to a minute and then plug it back in and let it reconnect to your internet provider.

If you use the Samsung Smart Switch to transfer your Bluetooth and Wi-Fi settings from an older device to the new one, it can cause problems with the connectivity of either

Bluetooth or Wi-Fi signals. To resolve this issue, navigate to **Settings**, then **General Management**, select **Reset**, then **Reset network settings**. This will clear out any problems from the data transfer, but you may have to set up new Bluetooth or Wi-Fi connections.

Dark Slow Motion Videos

One of the best features of the S9 is the impressive ability to film super slow motion video at a very high frame rate. This produces excellent videos, however, some videos come out too dark. This occurs because the video is underexposed and it can make it difficult to see details in the video. You can resolve this issue by disabling Start Stay on the S9 Plus. To disable it, first navigate to **Settings**, click on **Advanced features** and disable **Smart Stay**.

Dead Zone on Screen

Some Galaxy S9/S9 Plus has issues with dead zones on the phone. Dead zones are areas of the display that do not respond to your touch. Sometimes, an entire part of a display stops working.

The first step in diagnosing a dead zone is to make sure that there really is one. You can do this by opening the hardware diagnostics page. You can enter this mode by opening the phone dialer and dialing *#0*#. Access the Touch option and swipe your finger across the sections to discover if there is an area that is unresponsive to your touch. If there is a dead zone, you should contact Samsung for a replacement or repair.

The dead zone may not actually be dead. In cases like this, you may need to enable touch sensitivity. The Touch sensitivity option was discussed in Chapter 2.

Edge Lighting Not Working

Edge Lighting is a feature of the phone where it keeps certain information on the screen even when the phone is in sleep mode or the screen is off. The Apps edge lets you see recent apps, Tasks allows you to set Edge Lighting with common activities like sending a text message or adding a calendar event. It also shows notifications by lighting up when you are receiving calls or notifications, even when the phone is face down.

When Edge Lighting is not working, the screen does not light up for calls or notifications and the feature only appear to work with some apps.

If Edge Lighting is not working, there are a few tricks to try to turn it on. For instance, enabling pop-up notifications for apps that frequently use notifications such as Snapchat.

Another way you can resolve this problem is by enabling Animation Duration Scale. You can access this by navigating to **Settings**, locate **Developer Options**, and click on **Animation duration scale**. Use the slider to turn the feature on. You should set it to the 0.5x setting.

In order to enable Animation Duration Scale, you need to access the Developer Options. To access the Developer Options, navigate to **Settings**, click on **About Phone**, and select **Software Information**. On that screen, tap the **Build** number until you get a pop-up that states **Developer mode has been enabled**.

Another way to resolve the Edge Lighting issue is to download the Edge Lightning App from the Google Play Store. The app allows you to set custom colors for apps and helps the Edge Lighting to work even when the screen is off.

Keyboard not Appearing

Some users note that after they log into the phone with a PIN or password, the keyboard does not appear where it is supposed to. To resolve this problem, navigate to **Settings**, select **Apps**, tap the **three vertical dots**, then tap on **Show system settings**. Next, locate **Advanced Settings** and enable **Apps that appear on the top**. It should be noted that this permission is needed even if you are using a different keyboard such as Gboard.

Messaging App Sending Photos Automatically

A known issue with the S9 Plus is that it automatically sends photos via the Messaging app. These photos go out to people in your contacts. This bug brings up privacy concerns because you might not want to share your photos with certain people from your contact list. Samsung released an over the air (OTA) update for the problem. If the update does not help or is not installing correctly, you can also manually fix the problem. First, you need to navigate to the **Messaging App**. Click on **Permissions** and **turn storage off**. Once the app no longer has permission to access the storage, it cannot send the photos.

Missing App Drawer

Most Android phones have the app drawer on the Home screen, but Samsung phones do not. Having the App Drawer on the Home screen is a very easy way to access it but it takes up screen real-estate. If you like the app drawer on the Home screen, there is a way to put it back. First, press and hold on the Home screen. This will bring up the **layout editor**. Click on the **gear icon** and find the **Apps Button**. Next, swipe to enable the **show apps button**.

Out of Storage

One problem that you may experience with your S9/S9 Plus is running out of storage. Running out of space can be a large issue, especially if you are shooting a lot of videos. Here are some methods for freeing up space on your S9/S9 Plus. The phone has a slot for an SD card. SD cards range in storage size from 32GB cards to 512GB cards. The S9/S9 Plus has either 64GB or 128GB hard drive, so an SD card can more than double your storage space.

Once you have an SD card in your phone, you can move apps and other files to the SD card. You can do this with apps by navigating to **Settings**, scroll down to **Apps**. Choose the app you want to move and tap **App info**. Click on **Storage** and press **Change** to move to the app to the SD Card. In order to transfer other files to the SD card, swipe down from the Home screen and navigate to **Apps**, then click the **Samsung folder** and select **My Files**. Locate the file you want to move and touch **Move** or **Copy**, and then select the SD card. Navigate to where you want to save the file and touch **Done**.

You can also set the SD card as the default storage for the phone, rather than having to move the files manually to the SD card. In order to make this change, you need to access the **Developer Options**.

Once you make sure that Developer Options is enabled, navigate to **Settings**, and scroll down until you find **Developer Options**. Once you have found it, select it. Locate **Force allow apps on external** and enable it by swiping the slider into the on position. Next, you need to exit Settings and power off your phone and power it back on so that the changes take effect.

Another way to address the storage on your device is to use cloud storage services like Google Photos, Dropbox, or Microsoft One Drive. You can choose to upload files to the cloud service. This is a very useful trick since it also lets you access the files across your devices including any computers you might have.

Syncios Manager, which was described in Chapter 1 for transferring data between devices, is also impressive file management software. Using this software, you can control which files go into which folder on the phone.

Finally, you can free up storage by doing a factory reset, which removes all the files and settings on the phone. This includes deleting unnecessary files that are clogging up the phone's storage. Instructions on how to perform a factory reset are found in Chapter 5.

Samsung Pay

Samsung Pay allows you to pay for goods and services with your cell phone. Going cashless is a modern connivance and having your payment information in your phone allows you to keep your cash at home. The S9/S9 Plus has a problem with Samsung Pay where the app crashes too frequently and it prompts for an update even though it is the most recent version. There is a software update that solves this issue but if the issue persists, you may have to deal with it by manually installing the update. You can resolve it by navigating to the **Galaxy App Store**. The Samsung Pay app update will be listed here. Install the update, and it should resolve the issue with Samsung Pay.

Takes Too Long to Charge

The biggest problem that cell phones face is not having

enough power. Sometimes, it takes a too long to charge a phone from empty to full. In addition to the methods of power saving from the previous chapter, there is another method of increasing battery life by using fast charging. Unlike the other methods, this one is about how fast the phone charges and not how to save the battery life. The S9 Plus has an option for faster charging, which cuts down on the amount of time a phone has to spend on the charger. In order to activate the fast charging option, navigate to **Settings**, click on **Battery** and make sure that **Fast Charging** is enabled.

Video Stutters When Recording in 4k

Some videos shot in 4k resolution stutter while filming or viewing the video. There are two solutions to this problem. The first is to check the speed of the SD card you are using. The SD card may not be fast enough to deal with the data from the 4k video. To improve this, you will need to purchase a new and faster SD card. If you check the SD card and that does not affect the video stuttering, you may need to turn off electronic image stabilization or EIS. To disable EIS, navigate to the **Camera** app, open **Settings**, scroll until you locate **EIS**, and finally disable the **EIS**. Another setting that helps with 4k video is HEVC (High-efficiency video coding). You can enable HEVC in the same menu as disabling EIS.

Wet Speaker Audio Problems

After the S9/S9 Plus gets wet from being submerged in water, the wet speaker can affect the quality of the sound from the device. For the most part, the S9 Plus is waterproof. However, you might still experience problems with the speaker. Generally, this audio problem is due to software settings rather than physical damage to the phone.

There are reports that a wet speaker causes audio problems like crackling sounds, quiet audio, and not being able to hear callers unless being on speaker phone.

After the phone is found in water, it needs to be powered off and should be wiped off with a towel, then allowed to fully dry. After the phone is dry, test the speaker to see if the sound is degraded. If it is, load Safe Mode to see if the problem persists.

To load Safe Mode, press and hold the **Power** button. The Safe Mode prompt will appear. Tap the prompt and the phone will boot into Safe Mode. Safe Mode allows you to test apps to see if they are working properly.

If the problem persists, the other option is to back up the phone and do a factory reset (see Chapter 5). Once you have done the reset, test the phone to see if the problem persists. If the problem persists, there might be water damage to the hardware of the phone. If there is physical damage, you may need to contact a service center or take the phone in for repairs.

Chapter 5: S9/S9 Plus Maintenance

It is important to perform maintenance on your phone regularly to make sure the phone stays in good working order. Below are a variety of tools and steps to perform maintenance on your S9/S9 Plus.

Antivirus

Having an antivirus app is very important. There are many viruses and malware that target Android phones and some target Samsung phones in particular. To protect yourself, you should have an antivirus app installed on your phone. In Chapter 3, we discussed the app CleanMaster, however, there are many antivirus apps available on the Google Play store. Notable examples of other antivirus apps are Norton Security Antivirus, Avast Antivirus and Security, and McAfee Mobile Security. Each app offers several tools in addition to antivirus and malware removal.

Battery

Elsewhere in the book, we discussed Battery issues and how to conserve battery power. Here is a variety of tools and settings that allow you to monitor your battery usage, perform maintenance, and maintain your battery. To access the tools, navigate to **Settings**, select **Device Maintenance**, and tap **Battery**. You will get four options: Battery usage, Power saving mode, app power monitor, and More Options. The last is where you can access Advanced settings for the battery.

The **Battery usage** option allows you to view the details of your battery usage. Here, you can see what is using the battery and how much power each item is using.

Power saving mode allows you to extend the life of your battery. Under power saving mode, you can choose from Off, Mid, or Max. Each one allocates the battery usage differently. Power saving mode will give you an estimate on how long the battery will last in each of the power saving modes.

App power monitor allows you to prevent apps from using the battery by putting them to sleep when you are not using them.

Finally, you have **Advanced settings**. The Advanced settings allow you to modify the advanced battery options.

Backup

One crucial step in device maintenance is to backup your phone. This is important in case you accidentally delete files or have to do a hard reset or a master reset. These resets will remove all content and settings from the phone, and if you do not have a backup, you will have to adjust all of those settings again, as well as potentially losing data.

There are several storage options for backing up your phone. There are cloud storage options such as Google backup or Samsung Cloud. You can also save a backup onto your computer with Syncios Data Transfer, which was discussed in Chapter 1.

Backing up your data with cloud-based storage services is the same no matter which storage option you are using. In order to backup to a cloud service, navigate to **Settings**, click on **Cloud and accounts**, select the cloud service you want to use, and tap **Back up my data**. Use the sliders to select which content you want to be backed up then press **Back up now**. The phone will backup to the service you selected. For additional protection, you can back the phone up with multiple cloud services in order to protect your data from corruption.

The next way to back up your phone is to use a program on your computer to create the backup. Syncios Data Transfer is a program that can create backups. To create a backup in Syncios, first, select the **Backup** mode on the main screen and connect your phone with a USB cable. You can then customize where on the computer the backup will be stored. Once you have chosen where to store your backup, press **Next**. The following screen will allow you to pick the content you want to backup.

Press **Next** and the files will transfer to your computer. Once completed, you can restore from the backup using Syncios.

Device Maintenance App

Samsung released a Device Maintenance App for the Google Play store. This app helps you to perform maintenance on your phone and keep it working. The app is designed for any Samsung phone but works quite well with the S9/S9 Plus. The Device Maintenance App has the following features:

- Gives the phone a score between 1 and 100 depending on how the device is running.

- Performs optimization with a single click.

- Analyses the battery usage and creates a report of which apps are using power, and saves battery life by halting unused programs running in the background.

- Locates apps that are draining the battery.

- Features Power saving mode and Maximum power saving mode to extend the life of the battery.

- Allows you to remove unnecessary files
- Select the desired Performance mode

- Frees up RAM

- Detect viruses and malware and gives real-time protection against such intrusions.

Reset

The S9 has three different ways to reset the phone. The first is a soft reset, the second is a hard reset or factory reset, and the third is a master reset. A soft reset closes the apps being used and restarts the phone. This is essentially a way to restart the phone without unnecessary apps being open. A hard reset takes the phone back to factory settings, which will delete all files and settings on the phone. It is very important to have a backup before doing a hard reset. A master reset also deletes all files and settings. The master reset allows you to reset the phone if you are not able to access Settings or if the phone does not boot up properly.

To perform a soft reset on the S9/S9 Plus, press and hold the **Power** and **Volume down** buttons for roughly 10 seconds. The phone will then restart.

You can also do a hard reset from Settings. To perform the reset, navigate to **Settings**, tap **General management**, then press **Reset**, select **Factory data reset**, press **Reset**, and press **Delete all**.

To do a master reset, press the **Bixby** and the **Volume up** buttons. While you are pressing both buttons, press the **Power** button to restart the phone. Once the Samsung logo appears, release all of the buttons. Another screen will appear with an **Installing system update** and **no command** for roughly 30 seconds. Next, the **Android recovery menu** will appear. Use the up and down volume buttons to select **Wipe data / factory reset**. Press **Power** to select your option. The phone will then erase your data and restart with factory settings.

Memory

If your phone becomes slow and has other bugs, you may need to clean your memory. Cleaning your memory clears background apps and processes from the RAM of your phone, which can speed it up significantly. In order to clean the memory, navigate to **Settings**, select **Device maintenance**, touch **Memory**, and tap **Clean Now**.

Performance

You can control which Performance mode you want to use. These performance modes are designed with different functions in mind. The Performance mode options are optimized, game, entertainment, and high performance. The Optimized mode is a good balance between screen resolution and the battery life of the device. The optimized mode is the recommended mode for general use. As the name indicates, the game mode is good for playing games on your phone. Likewise, entertainment mode is good for viewing pictures and listening to music. The entertainment mode features high-quality sounds and crisp videos. Finally, high-performance mode features the highest display settings for viewing videos in a very high resolution.

Conclusion

Thank you for making it through to the end of ***Ultimate Samsung Galaxy S9 & S9 Plus Manual: Tips, Tricks, and Great Optimization for Your Samsung Galaxy S9***. Let's hope it was informative and able to provide you with all of the tools you need to achieve your goals whatever they may be. Now that you have finished this book, the opportunities are endless, but there is still much to learn. Keep reading and learning more about your Samsung S9 or S9 Plus to broaden your perspective. Learning more is the only way to master the Galaxy S9 skills you need in order to take control of your mobile life and use your phone like a pro.

The next step is to take out your phone and begin customizing and optimizing it. Make the tips and tricks you learned throughout this book work for you. Use your newfound skills to download the best apps, backup your phone in case of emergencies, play games, watch videos, and take videos in amazing 4k resolution. Put your knowledge to the test against any problem you might experience with your phone, but don't stop there. Take your new phone skills to the max.

Finally, if you found this book useful in any way, a review on Amazon is always appreciated!

www.ingramcontent.com/pod-product-compliance
Lightning Source LLC
Chambersburg PA
CBHW030507220526
45464CB00006B/2695